INTERNATIONAL CENTRE FOR MECHANICAL SCIENCES

COURSES AND LECTURES - No. 58

HEINZ PARKUS
TECHNICAL UNIVERSITY OF VIENNA

VARIATIONAL PRINCIPLES IN THERMO- AND MAGNETO-ELASTICITY

COURSE HELD AT THE DEPARTMENT
FOR MECHANICS OF DEFORMABLE BODIES
OCTOBER 1970

UDINE 1970

SPRINGER-VERLAG WIEN GMBH

This work is subject to copyright.

All rights are reserved,

whether the whole or part of the material is concerned

specifically those of translation, reprinting, re-use of illustrations,

broadcasting, reproduction by photocopying machine

or similar means, and storage in data banks.

© 1972 by Springer-Verlag Wien
Originally published by Springer-Verlag Wien - New York in 1972

ISBN 978-3-211-81080-4   ISBN 978-3-7091-2941-8 (eBook)
DOI 10.1007/978-3-7091-2941-8

PREFACE

This short monograph is intented as a textbook for a series of lectures given by the author at the Centre International des Sciences Mécaniques in Udine in the fall of 1970.

While applications of the calculus of variations to problems of continuum mechanics have a long and fascinating history interest in the special subjects treated in this monograph is relatively new. I hope, therefore, that the attempt undertaken here, at a unified presentation of the material scattered over a large number of scientific journals might serve a useful purpose.

The limited time available for the lectures did not permit for the treatment of specific applications. A few references to the pertinent literature are given instead.

Some familiarity from the part of the reader with the concepts and techniques of the calculus of variations is expected. All mathematical subtleties, however, have been omitted. A certain knowledge of the fundamentals of thermodynamics and electrodynamics will be necessary for an understanding of some parts of the text. I take great pleasure in expressing my sincere gratitude to the Secretary General of CISM, Prof. L. Sobrero, and to the Rector, Prof. W. Olszak, for inviting me to present these lectures.

Parkus

Udine, October 1970

CONTENTS

	Page
Preface	3
Introduction	5
Chapter I. Uncoupled Thermoelasticity	7
Chapter II. Coupled Thermoelasticity	11
(1) Generalized Hamilton's Principle	11
(2) Herrmann's combined Biot-Reissner Principle	16
Chapter III. Thermo-Viscoelasticity	22
(1) The principle by Sanders et al	23
(2) The principle by Olszak and Perezyna	27
(3) Schapery's Principle	28
Chapter IV. Magnetoelasticity	32
Chapter V. Piezoelectricity	41
References	45

INTRODUCTION

Recognition of the fact that boundary value problems of continuum mechanics are equivalent to problems of the calculus of variations goes back to Daniel Bernoulli [1] who considered the special case of the elastic rod. The formulation of the general three-dimensional elastic problem has been given by Green [2] for the displacements and by Castigliano [3] for the stresses. The extension to deformable bodies of Hamilton's principle is due to Kirchhoff [4].

In recent years the classical variational principles of elasticity have been extented in various ways to include viscoelasticity, thermal stress, heat conduction and magnetic and electric effects. These extensions will be discussed in the present short monograph. Solids with microstructure, however, will not be included. Also, variational principles in the theory of plasticity have been omitted since these will be treated in a parallel course.

It is well known that variational principles can serve many purposes. First, they may be used to derive consistent sets of differential equations and boundary conditions. The classical example here is Kirchhoff's treatment of the transverse bending of thin plates. Second, they may be employed to prove existence and uniqueness of the corresponding solutions. Third, different variational principles applied simultaneously may be used to establish upper and lower bounds for certain quantities like displacements or stresses. Finally, they may be taken as the starting point for the development of approximation methods and numerical procedures. Best known of these is the Ritz method.

It is the principle of virtual displacements on the one hand and the principle of virtual forces on the other that form the foundation of all purely mechanical variational principles. By direct application of the former (or its generalization to dynamics, the Alembert's principle) to conservative systems Green's principle of minimum potential energy is immediately obtained. Similarly, application of the principle of virtual forces renders Castigliano's principle of minimum complementary energy. However, this situation changes, if mixed principles like those by E. Reissner [5] are considered, or if nonmechanical effects, e.g. from thermodynamics or magnetodynamics are included. A frequently used procedure for obtaining the corresponding variational principles is then to guess at their form and to show subsequently that they render the correct differential equations and boundary conditions of the problem which, of course, have to be known already. This - admittedly unsatisfactory - procedure will, of necessity, be followed in most parts of this monograph. Notable exceptions are chapters 4 and 5.

CHAPTER I
UNCOUPLED THERMOELASTICITY

In the uncoupled thermoelastic case the equation of heat conduction separates from the remaining equations*). It is, therefore, an easy matter to generalize the principles of isothermal elasticity to this case. In fact, all that has to be done is to include in the expression for the elastic potential the dependence on temperature**):

$$W = W(\varepsilon_{ij}, \theta), \quad W^* = W^*(\tau_{ij}, \theta). \tag{1.1}$$

Here, $\theta = T - T_0$ is the temperature excess above an absolute reference temperature T_0. In particular, if Hooke's law is valid,

$$W^* = \frac{1}{2} b_{ijkl} \tau_{ij} \tau_{kl} - \gamma_{ij} \tau_{ij} \theta + \frac{c \theta^2}{2 T_0}. \tag{1.2}$$

The material constants obey the symmetry properties

$$b_{ijkl} = b_{jikl} = b_{klij}, \quad \gamma_{ij} = \gamma_{ji} \tag{1.3}$$

---

*) see, for instance, [7], Sec. 2.2.
**) [7], Sec. 5.5.

and reduce to

(1.4) $$b_{ijkl} = B\delta_{ik}\delta_{jl} + C\delta_{ij}\delta_{kl}, \quad \gamma_{ij} = \gamma\delta_{ij}$$

for the isotropic solid.

For instance, Reissner's mixed principle [5] may now be stated as

(1.5) $$\delta I = 0$$

where

(1.6) $$I = \int_V [\varepsilon_{ij}\tau_{ij} - F_i u_i - W^*]dV - \int_{A_1} \bar{p}_i u_i dA - \int_{A_2} p_i(u_i - \bar{u}_i)dA$$

and

(1.7) $$\varepsilon_{ij} = \tfrac{1}{2}(u_{i,j} + u_{j,i}),$$

u_i are the displacement components, F_i the body forces and p_i the surface stresses. A bar indicates prescribed values. Thus, $\bar{p}_i$ are the stress components prescribed on part A_1 of the surface A of the body, and $\bar{u}_i$ are the displacements prescribed on part $A_2 = A - A_1$ of the surface.

The quantities τ_{ij} and u_i are varied <u>independently</u> while temperature θ is to be kept constant:

$$\delta I = \int_V \left[\varepsilon_{ij}\delta\tau_{ij} + \tau_{ij}\delta\varepsilon_{ij} - F_i\delta u_i - \frac{\partial W^*}{\partial \tau_{ij}}\delta\tau_{ij}\right]dV -$$

$$-\int_{A_1} \bar{p}_i \delta u_i dA - \int_{A_2} (u_i - \bar{u}_i) \delta p_i dA .$$

The second term in the volume integral is now transformed by integration by parts, using (1 - 7) and $p_i = \tau_{ij} n_j$,

$$\frac{1}{2}\int_V \tau_{ij} \delta(u_{i,j} + u_{j,i}) dV = \int_V [(\tau_{ij} \delta u_i)_{,j} - \tau_{ij,j} \delta u_i] dV =$$

$$= \oint_A p_i \delta u_i dA - \int_V \tau_{ij,j} \delta u_i dV = \int_{A_1} p_i \delta u_i dA - \int_V \tau_{ij,j} \delta u_i dV$$

since $\delta u_i = 0$ on A_2. Substitution into δI gives

$$\delta I = \int_V \left[\left(\varepsilon_{ij} - \frac{\partial W^*}{\partial \tau_{ij}}\right) \delta \tau_{ij} - (\tau_{ij,j} + F_i) \delta u_i \right] dV + \\ + \int_{A_1} (p_i - \bar{p}_i) \delta u_i dA - \int_{A_2} (u_i - \bar{u}_i) \delta p_i dA = 0 \quad (1.8)$$

and this yields the Euler equations

$$\varepsilon_{ij} = \frac{\partial W^*}{\partial \tau_{ij}} , \quad \tau_{ij,j} + F_i = 0 \quad \text{in } V \quad (1.9)$$

and

$$p_i = \bar{p}_i \quad \text{on } A_1 , \quad u_i = \bar{u}_i \quad \text{on } A_2 . \quad (1.10)$$

Green's variational principle for the displacements and Castigliano's variational principle for stresses follow from Reissner's principle as special cases. This will be discussed in a more general setting in the next chapter.

What is still needed is a variational principle for the equation of heat conduction. This will also given in the next chapter.

Uncoupled variational principles have been used by <u>Trostel</u> [8] to obtain approximate solutions to thermal stress problems.

CHAPTER II
COUPLED THERMOELASTICITY

Variational principles for coupled thermoelasticity have first been established by Parkus [9] in the form of a generalized Hamilton's principle, and by Biot [10]. Herrmann [11] has applied Biot's ideas to Reissner's variational principle. Later on, other principles have been formulated by Ben-Amoz [12], Nickell and Sackman [13] and Rafalski [14], among others.

(1) Generalized Hamilton's Principle. We consider the general dynamic case of nonlinear thermoelasticity. Let

$$K = \frac{1}{2} \int_m \dot{u}_i \dot{u}_i \, dm \qquad (2.1)$$

be the kinetic energy of the moving body of mass m and define two functionals Π and Φ as

$$\Pi = \int_m (F + ST - F_i u_i) dm + \oint_A P_i u_i \, dA \qquad (2.2)$$

$$\Phi = \int_m (H - ST\dot{T} - RT) dm + \oint_A Q_i T n_i \, dA. \qquad (2.3)$$

Here $F(e_{\alpha\beta}, T)$ is the free energy per unit mass, S is the entropy per unit mass, T is absolute temperature. P_i and Q_i are surface stress and heat flux vector, respectively, referred to the unit of the deformed surface. n_i is the unit normal vector on the surface, positive outward. R is the heat produced per unit of time and unit of mass by heat sources distributed in the body. H is a "heat flux potential" per unit mass, defined by

$$(2.4) \quad q_\alpha = -\varrho_0 \frac{\partial H}{\partial T_{,\alpha}}, \quad \varrho_0 H = \frac{1}{2} a_{\alpha\beta} T_{,\alpha} T_{,\beta}$$

where $q_i = Q_i \sqrt{g}$ represents the heat flux vector in the deformed body referred, however, to the unit surface in the undeformed (initial) state. ϱ_0 is the mass density in this state. The components e_{ij} of the Green strain tensor are defined by

$$(2.5) \quad 2e_{\alpha\beta} = g_{\alpha\beta} - \delta_{\alpha\beta} = x_{i,\alpha} x_{i,\beta} - \delta_{\alpha\beta} = u_{i,\alpha} \delta_{i\beta} + u_{i,\beta} \delta_{i\alpha} + u_{i,\alpha} u_{i,\beta}.$$

All quantities are assumed functions of the initial coordinates X_i and time t. The usual notation $\partial u_m / \partial X_\alpha = u_{m,\alpha}$ is used. The functions

$$x_i = X_i + u_i$$

are more convenient in finite-displacement theory than are the displacements.

The principle may now be stated as follows. Consider two arbitrary time instants t_1 and t_2. Then, for all admissible variations of the motion of the body,

$$(2.6) \quad \delta \int_{t_1}^{t_2} (K - \Pi) dt = 0 \quad \text{and} \quad \delta \int_{t_1}^{t_2} \Phi \, dt = 0.$$

Admissible motions satisfy the stress-strain relations, are compatible with the kinematical constraints on the body and, in addition, coincide with the actual motion at time t_1 and t_2. Displacements u_i and temperature T are varied while time t, external forces F_i and P_i, heat sources R, heat flux Q_i and entropy S are kept unchanged.

For a proof one has

$$\delta \int_{t_1}^{t_2} K dt = \int_m \int_{t_1}^{t_2} \dot{u}_i \frac{d\delta u_i}{dt} dt\, dm = \int_m \left[\dot{u}_i \delta u_i \Big|_{t_1}^{t_2} - \int_{t_1}^{t_2} \ddot{u}_i \delta u_i dt \right] dm . \quad (2.7)$$

The first term vanishes. Furthermore

$$\delta \int_{t_1}^{t_2} \Pi dt = \int_{t_1}^{t_2} dt \left\{ \int_m \left[\frac{\partial F}{\partial e_{\alpha\beta}} \delta e_{\alpha\beta} + \frac{\partial F}{\partial T} \delta T + S \delta T - F_i \delta u_i \right] dm - \oint_A P_i \delta u_i dA \right\}.$$

But, from Eq. (2.5),

$$2\delta e_{\alpha\beta} = (\delta_{m\alpha} + u_{m,\alpha}) \frac{\partial}{\partial X_\beta} \delta u_m + (\delta_{m\beta} + u_{m,\beta}) \frac{\partial}{\partial X_\alpha} \delta u_m .$$

Making use of the stress-stain relation*)

$$S_{\alpha\beta} = \varrho_0 \frac{\partial F}{\partial e_{\alpha\beta}} \quad (2.8)$$

---

*) see, for instance, [7], Sec. 5.3.

where $S_{\alpha\beta} = S_{\beta\alpha}$ is the second Piola-Kirchhoff stress tensor, writing $dm = \varrho_0 dV_0$ and introducing the deformation tensor

$$(2.9) \qquad f_{m\alpha} = \delta_{m\alpha} + u_{m,\alpha}$$

one obtains therefore

$$\int_m \frac{\partial F}{\partial e_{\alpha\beta}} \delta e_{\alpha\beta} dm = \int_{V_0} S_{\alpha\beta} f_{m\alpha} \frac{\partial}{\partial X_\beta}(\delta u_m) dV_0 =$$

$$= \int_{V_0} \frac{\partial}{\partial X_\beta}(f_{m\alpha} S_{\alpha\beta} \delta u_m) dV_0 - \int_{V_0} \frac{\partial}{\partial X_\beta}(f_{m\alpha} S_{\alpha\beta}) \delta u_m dV_0 =$$

$$= \oint_{A_0} f_{m\alpha} S_{\alpha\beta} n_\beta^0 \delta u_m dA_0 - \int_{V_0} (f_{m\alpha} S_{\alpha\beta})_{,\beta} \delta u_m dV_0 .$$

Hence

$$(2.10) \quad \delta \int_{t_1}^{t_2} (K - \Pi) dt = \int_{t_1}^{t_2} dt \Big\{ \int_{V_0} \Big[(f_{m\alpha} S_{\alpha\beta})_{,\beta} + \varrho_0 F_m - \varrho_0 \ddot{u}_m \Big] \delta u_m -$$

$$- \int_m \Big[\frac{\partial F}{\partial T} + S\Big] \delta T dm + \oint_{A_0} \Big[p_m - f_{m\alpha} S_{\alpha\beta} n_\beta^0 \Big] \delta u_m dA_0 \Big\} = 0 ,$$

where we have put

$$(2.11) \qquad P_i dA = p_i dA_0$$

and p_i represents the surface load on the deformed body referred, however, to the unit of the undeformed surface. n_β^0 is the normal vector of the undeformed surface.

From Eq. (2 - 10) we conclude on the basis of the fundamental lemma that the following equations and boundary conditions hold:

$$\left. \begin{array}{l} (f_{m\alpha} S_{\alpha\beta})_{,\beta} + \varrho_0 F_m = \varrho_0 \ddot{u}_m \\ \\ S = -\dfrac{\partial F}{\partial T} \end{array} \right\} \quad \text{everywhere in the body} \quad (2.12)$$

$$f_{m\alpha} S_{\alpha\beta} n_\beta^0 = p_m \quad \text{on that part of the surface where the displacements are not prescribed.} \quad (2.13)$$

Similarly, from the second of Eqs. (2 - 6)

$$\delta \int_{t_1}^{t_2} \Phi \, dt = \int_{t_1}^{t_2} dt \left\{ \int_m \left[\dfrac{\partial H}{\partial T_{,\alpha}} \dfrac{\partial}{\partial X_\alpha} \delta T - S\dot{T}\delta T - ST\dfrac{\partial}{\partial t}\delta T - R\delta T \right] dm + \oint_A Q_i n_i \delta T dA \right\} =$$

$$= \int_{t_1}^{t_2} dt \left\{ \int_{V_0} [q_{\alpha,\alpha} - \varrho_0(R - T\dot{S})] \delta T \, dV_0 - \int_{V_0} (q_\alpha \delta T)_{,\alpha} \, dV_0 + \oint_A Q_i n_i \delta T dA \right\}.$$

But, from Gauss' theorem, using $\sqrt{g} \, dV_0 = dV$,

$$\int_{V_0} (q_\alpha \delta t)_{,\alpha} \, dV_0 = \int_V \dfrac{1}{\sqrt{g}} \left(\dfrac{q_\alpha}{\sqrt{g}} \delta T \sqrt{g} \right)_{,\alpha} dV = \oint_A \dfrac{q_i}{\sqrt{g}} \delta T n_i dA .$$

The following equation of heat conduction and boundary condition are therefore obtained:

(2.14) $\quad q_{\alpha,\alpha} = -(a_{\alpha\beta}T_{,\beta})_{\alpha} = \varrho_0(R - T\dot{S})$ in the body

(2.15) $\quad\quad\quad q_i = \sqrt{g}\, Q_i$ on the surface

with

(2.16) $\quad\quad\quad \sqrt{g} = \dfrac{\varrho_0}{\varrho}.$

In the stationary (static) case Eqs. (2.6) reduce to

(2.17) $\quad\quad\quad \delta\Pi = 0$ and $\delta\Phi = 0$

and the equation of heat conduction (2 - 14) separates from Eqs. (2 - 12).

In performing the variation care has to taken to keep S constant (isentropic variation). For that same reason $F + ST$ in Eq. (2.2) must not be replaced by the internal energy U.

(2) Herrmann's combined Biot-Reissner Principle

[11] . This principle is restricted to linear, quasistatic thermoelasticity. Let

$$I = \int_V (\varepsilon_{ij}\tau'_{ij} - F_i u_i - W^* - \gamma\theta - s_i g_i - D)dV - $$

$$-\int_{A_1}\bar{p}_i u_i dA - \int_{A_2} p_i(u_i - \bar{u}_i)dA +$$

$$+\int_{A_3}\theta(s_i - \bar{s}_i)n_i dA + \int_{A_4}\bar{\theta} s_i n_i dA .$$

(2.18)

Then

$$\delta I = 0 \qquad (2.19)$$

where the functions in each product have to be varied independently.

Apart from the terms already used in Chapter 1 the notation is as follows. The vector s_i is Biot's "entropy displacement", defined as

$$s_{i,i} = -S . \qquad (2.20)$$

For small temperature changes s_i is, approximately, proportional to the amount of heat which has flown in a given direction :

$$s_i = \frac{1}{T_0}\int_0^t q_i dt . \qquad (2.21)$$

This follows from $dS = dh/T$ or, approximately, with $\dot{h} + q_{i,i} = 0$, from

$$S = \frac{h}{T_0} = \frac{-1}{T_0}\int_0^t q_{i,i} dt = -s_{i,i}$$

h is the amount of heat absorbed by the unit volume.

The quantity γ denotes the "thermoelastic dilatation",

(2.22) $$\gamma = s_{i,i} + \beta_{ij} w_{i,j}$$

where β_{ij} are the coefficients of thermal dilatation in the equation of equilibrium

(2.23) $$\tau'_{ij,j} + F_i = \beta_{ij} g_j$$

and τ'_{ij} is only that part of the stress which is not due to the temperature θ, and g_i is the temperature gradient,

(2.24) $$g_i = \theta_{,i}$$

to be treated as an independent variable. D is Biot's dissipation function defined as

(2.25) $$\dot{D} = \frac{a_{ij}}{T_0} g_i g_j$$

cf. Eq. (2.4). From Fourier's law

$$\dot{D}_{,g_i} = -\frac{q_i}{T_0}$$

it follows, using Eq. (2.21),

(2.26) $$s_i = -\dot{D}_{,g_i} .$$

Eqs. (2.22) and (2.26), together with the relation

(2.27) $$W^*_{,\theta} = -\gamma$$

are equivalent to the equation of heat conduction, and replace it.

The proof of (2.19) is along the lines given in the preceding principle and in Chapter 1. Eqs. (2.23), (2.24), (2.27) and the first of Eqs. (1.9), with τ_{ij} replaced by τ'_{ij}, are obtained as Euler equations while the two equations (1.10) together with

$$s_i = \bar{s}_i \quad \text{on part } A_3$$

$$\text{of the surface } A = A_3 + A_4 \quad (2.28)$$

$$\theta = \bar{\theta} \quad \text{on part } A_4$$

follow as the corresponding boundary conditions. s_i and θ are to be prescribed on A_3 and A_4, respectively.

The practical application of this principle, in particular in connection with the Ritz method, appears to be restricted due to the presence of the somewhat abstract quantity s_i for which it should be difficult to find a good Ritz representation.

By different choices of functionals and suitable restrictions on the variations the principles of Green and Castigliano generalized to coupled thermoelasticity may be obtained. Indeed, introducing two constitutive functions $W(\epsilon_{ij}, \gamma)$ and $G(s_i)$ such that

$$\tau'_{ij} = \partial W / \partial \epsilon_{ij} \quad (2.29a)$$

(2.29b)
$$\theta = -W_{,T}$$
$$g_i = \partial G/\partial s_i$$

one finds that the variation

(2.30)
$$\delta I_u = 0$$

of

(2.31) $$I_u = \int_V (W - G - F_i u_i)dV - \int_{A_1} p_i u_i dA + \int_{A_4} \theta s_i n_i dA$$

yields Eqs. (2.23) and (2.24) and boundary conditions $(1.10)_1$ and $(2.28)_2$, provided the two vectors δu_i and δs_i are varied independently.

On the other hand, if in the functional

(2.32) $$I_\tau = \int_V (W^* + D - F_i u_i)dV - \int_{A_2} \bar{u}_i p_i dA + \int_{A_3} (\bar{s}_j + \beta_{ij} u_i) n_j \theta dA$$

with

(2.33)
$$\delta I_\tau = 0$$

the stresses τ'_{ij} and the temperature θ are varied independently but in such a way that Eqs. (2.23), (2.24) and (2.26) are always satisfied, the remaining two equations (1.9), and (2.27), together with boundary conditions $(1.10)_2$ and $(2.28)_1$ are recovered [15]. Briefly, the procedure is as follows.

$$\delta I_\tau = \int_V \left[\frac{\partial W^*}{\partial \tau'_{ij}} \delta \tau'_{ij} + \frac{\partial W^*}{\partial \theta} \delta\theta + \frac{\partial D}{\partial g_i} \frac{\partial(\delta\theta)}{\partial x_i} - u_i \delta F_i \right] dV -$$

$$- \int_{A_2} \bar{u}_i \delta p_i \, dA + \int_{A_3} (\bar{s}_i + \beta_{ij} u_i) n_j \delta\theta \, dA .$$

But, from Eq. (2.23), within the body,

$$\delta F_i = \beta_{ij} \frac{\partial}{\partial x_j} \delta\theta - \frac{\partial}{\partial x_j} \delta \tau'_{ij}$$

and, hence, since $\delta\theta = 0$ on $A - A_3$ and $\delta p_i = 0$ on $A - A_2$, assuming β_{ij} constant,

$$\delta I_\tau = \int_V \left[\frac{\partial W^*}{\partial \tau'_{ij}} \delta\tau'_{ij} + \frac{\partial W^*}{\partial \theta} \delta\theta - \frac{\partial}{\partial x_i}(D_{,g_i}) \delta\theta + \beta_{ij} u_{i,j} \delta\theta - \frac{1}{2}(u_{i,j} + u_{j,i}) \delta\tau'_{ij} \right] dV +$$

$$+ \int_{A_2} (u_i - \bar{u}_i) \delta p_i \, dA + \oint_{A_3} (\bar{s}_i - s_i) n_i \delta\theta \, dA = 0 .$$

This renders the equations mentioned above.

It should be pointed out that the situation concerning the type of stationarity of I, I_u and I_τ is precisely the same as in the isothermal case. The general Biot-Reissner variational theorem (2.19) is only a stationary - value problem while both I_u and I_τ take on minimum values.

Biot's principle in combination with the Ritz method has been used by Zeman[16] for the solution of stochastic linear thermoelastic problems.

CHAPTER III
THERMO-VISCOELASTICITY

Two classes of variational theorems in viscoelasticity may be distinguished. In the first, attention is focused on problems of secondary creep. Here, the quantities to be varied are stress rates or strain rates ; that is, if the state of stress and strain throughout the body is known at a given instant, then application of the variational theorem singles out the stress rates or strain rates (or, equivalently, the stress increments or strain increments) that actually occur from those rates of stress or strain that are admitted in the enunciation of the theorem*). A theorem of this type, based on work by Wang and Prager [17] , has been given by Sanders, McComb and Schlechte [18] for isothermal conditions. It will be extented below to include temperature effects. In the second class it is the stresses and strains themselves that are subject to variations.

Most of the work in this class is due to Biot [19]

---

*) The same situation occurs in variational theorems of the theory of plasticity.

and to Schapery [20]. The latter has advanced a variety of linearized principles for displacements, stress, entropy displacement and temperature. Less general theorems are due to Olszak and Perzyna [21] for the static case, and to Rafalski [22] for the dynamic case.

The well-known elastic analogy given by Alfrey, and its extension by Hoff to nonlinear viscoelasticity could, of course, also be used in a variational procedure, cf. [20] and [23].

(1) The Principle by Sanders et al. This principle is valid for finite deformations, eq. (2.5) and nonlinear stress strain relations. Static conditions are assumed and coupling between temperature and deformation is disregarded.

Let strain be separable into an elastic component $e^e_{\alpha\beta}$ and a creep component $e^c_{\alpha\beta}$,

$$e_{\alpha\beta} = e^e_{\alpha\beta} + e^c_{\alpha\beta} \qquad (3.1)$$

where the elastic component includes thermal strain:

$$e^e_{\alpha\beta} = \frac{\partial W^*(s_{\alpha\beta}, T)}{\partial s_{\alpha\beta}}. \qquad (3.2)$$

The creep strain rate $\dot{e}^c_{\alpha\beta}$ is supposed to be <u>independent of the stress rate</u>. This comprises practically all laws of secondary creep proposed so far.

The variational theorem then states that the integral

$$(3.3) \quad I = \int_{V_0}\left[\dot{e}_{\alpha\beta}\dot{s}_{\alpha\beta} + \frac{1}{2}\dot{u}_{k,\alpha}\dot{u}_{k,\beta}\dot{s}_{\alpha\beta} - \frac{1}{2}(\dot{e}^{\theta}_{\alpha\beta} + 2\dot{e}^{c}_{\alpha\beta})\dot{s}_{\alpha\beta}\right]dV_0 - \int_{A_1}\dot{u}_i\dot{\bar{p}}_i dA_0 - \int_{A_2}(\dot{u}_i - \dot{\bar{u}}_i)p_i dA_0$$

extented over the undeformed (initial) body takes on a stationary value

$$(3.4) \quad \delta I = 0$$

for the actual stress rates $\dot{s}_{\alpha\beta}$ and velocities $\dot{u}_i$. The load rate $\dot{\bar{p}}_i$ is prescribed on part A_1 of the surface, and the velocity $\dot{\bar{u}}_i$ is prescribed on part A_2 of the surface. $\dot{e}_{\alpha\beta}$ is understood to be written in terms of displacements and velocities as obtained by differentiating Eq. (2.5):

$$(3.5) \quad 2\dot{e}_{\alpha\beta} = \dot{u}_{i,\alpha}\delta_{i\beta} + \dot{u}_{i,\beta}\delta_{i\alpha} + \dot{u}_{k,\alpha}u_{k,\beta} + u_{k,\alpha}\dot{u}_{k,\beta}.$$

On the other hand, $\dot{e}^{e}_{\alpha\beta}$ is in terms of stresses and stress rates, cf. Eq. (3.2), and $\dot{e}^{c}_{\alpha\beta}$ is in terms of stress.

The variation of (3.3) is

$$\delta I = \int_{V_0}\left[\dot{e}_{\alpha\beta}\delta\dot{s}_{\alpha\beta} + \dot{s}_{\alpha\beta}\delta\dot{e}_{\alpha\beta} + s_{\alpha\beta}\dot{u}_{k,\alpha}\delta\dot{u}_{k,\beta} - \frac{1}{2}\dot{s}_{\alpha\beta}\delta\dot{e}^{e}_{\alpha\beta} - \frac{1}{2}(\dot{e}^{e}_{\alpha\beta} + 2\dot{e}^{c}_{\alpha\beta})\delta\dot{s}_{\alpha\beta}\right]dV_0 -$$

$$-\int_{A_1} \bar{\dot{p}}_i \delta \dot{u}_i dA_0 - \int_{A_2} (\dot{u}_i - \bar{\dot{u}}_i) \delta \dot{p}_i dA_0 \ . \tag{3.6}$$

The second term in the volume integral may be integrated by parts using Eqs. (3.5) and (2.9):

$$\int_{V_0} \dot{s}_{\alpha\beta} \delta \dot{e}_{\alpha\beta} dV_0 = \oint_{A_0} \dot{s}_{\alpha\beta} f_{k\alpha} n_\beta^0 \delta \dot{u}_k dA_0 - \int_{V_0} (f_{k\alpha} \dot{s}_{\alpha\beta})_{,\beta} \delta \dot{u}_k dV_0 \ . \tag{3.7}$$

The third term may also be integrated:

$$\int_{V_0} s_{\alpha\beta} \dot{u}_{k,\alpha} \delta \dot{u}_{k,\beta} dV_0 = \oint_{A_0} s_{\alpha\beta} n_\beta \dot{u}_{k,\alpha} \delta \dot{u}_k dA_0 - \int_{V_0} (s_{\alpha\beta} \dot{u}_{k,\alpha})_{,\beta} \delta \dot{u}_k dV_0 \ . \tag{3.8}$$

Now, since $e_{\alpha\beta}^e$ is independent of the stress rates, $\dot{e}_{\alpha\beta}^e$, is homogeneous of order one in $\dot{s}_{\alpha\beta}^e$ i.e.,

$$\frac{\partial \dot{e}_{\alpha\beta}^e}{\partial \dot{s}_{\lambda\mu}} \dot{s}_{\lambda\mu} = \dot{e}_{\alpha\beta}^e \ .$$

Hence,

$$\dot{s}_{\alpha\beta} \delta \dot{e}_{\alpha\beta}^e = \dot{s}_{\alpha\beta} \frac{\partial \dot{e}_{\alpha\beta}^e}{\partial \dot{s}_{\lambda\mu}} \delta \dot{s}_{\lambda\mu} = \dot{s}_{\alpha\beta} \frac{\partial \dot{e}_{\lambda\mu}^e}{\partial \dot{s}_{\alpha\beta}} \delta \dot{s}_{\lambda\mu} = \dot{e}_{\lambda\mu}^e \delta \dot{s}_{\lambda\mu} \ . \tag{3.9}$$

This follows from

$$\dot{e}_{\alpha\beta}^e = \frac{d}{dt} \frac{\partial W^*}{\partial s_{\alpha\beta}} = \frac{\partial}{\partial s_{\alpha\beta}} \left(\frac{\partial W^*}{\partial s_{\lambda\mu}} \dot{s}_{\lambda\mu} + \frac{\partial W^*}{\partial T} \dot{T} \right)$$

and therefore,

$$\frac{\partial \dot{e}^e_{\alpha\beta}}{\partial \dot{s}_{\lambda\mu}} = \frac{\partial^2 W^*}{\partial \dot{s}_{\alpha\beta} \partial \dot{s}_{\lambda\mu}} = \frac{\partial \dot{e}^e_{\lambda\mu}}{\partial \dot{s}_{\alpha\beta}}.$$

Substituting Eqs. (3.7), (3.8) and (3.9) into (3.6), we get

$$\delta I = \int_{V_0} \left[e_{\alpha\beta} \delta \dot{s}_{\alpha\beta} - (f_{k\alpha} \dot{s}_{\alpha\beta})_{,\beta} \delta \dot{u}_k - (s_{\alpha\beta} \dot{u}_{k,\alpha})_{,\beta} \delta \dot{u}_k - (\dot{e}^e_{\alpha\beta} + \dot{e}^c_{\alpha\beta}) \delta \dot{s}_{\alpha\beta} \right] dV_0 +$$

$$+ \oint_{A_0} \left[\dot{s}_{\alpha\beta} f_{k\alpha} + s_{\alpha\beta} \dot{u}_{k,\alpha} \right] n^0_\beta \delta \dot{u}_k dA_0 -$$

$$- \int_{A_1} \dot{\bar{p}}_i \delta \dot{u}_i dA_0 - \int_{A_2} (\dot{u}_i - \dot{\bar{u}}_i) \delta \dot{p}_i dA_0$$

or, using

$$\dot{f}_{k,\alpha} = \dot{u}_{k,\alpha}, \quad s_{\alpha\beta} f_{k\alpha} n^0_\beta = p_k,$$

$$(3.10) \begin{cases} \delta I = \int_{V_0} \left[(\dot{e}_{\alpha\beta} - \dot{e}^e_{\alpha\beta} - \dot{e}^c_{\alpha\beta}) \delta \dot{s}_{\alpha\beta} - \delta \dot{u}_k \frac{d}{dt}(f_{k\alpha} s_{\alpha\beta})_{,\beta} \right] dV_0 + \\ + \int_{A_1} (\dot{p}_i - \dot{\bar{p}}_i) \delta \dot{u}_i dA_0 - \int_{A_2} (\dot{u}_i - \dot{\bar{u}}_i) \delta \dot{p}_i dA_0 = 0. \end{cases}$$

Vanishing of the coefficients of $\delta \dot{s}_{\alpha\beta}$, $\delta \dot{u}_i$, $\delta \dot{p}_i$ reproduces Eqs. (3.1), (2.12) with $\varrho_0 = 0$, and boundary conditions (1.10).

So far the theorem has been used only for the study of isothermal creep problems. For nonisothermal problems it would be important to generalize it to include the temperature-dependence of the material viscosity.

(2) The Principle by Olszak and Perzyna. The principle is valid for a linear viscoelastic medium with temperature-independent properties:

$$\tau_{ij} = c_{ijkl}\varepsilon_{kl} - \beta_{ij}\theta \qquad (3.11)$$

where β_{ij} and c_{ijkl} are linear operators. $\theta = T - T_0$ and ε_{kl} is given by Eq. (1.7). Inverting relation (3.11) we obtain

$$\varepsilon_{ij} = b_{ijkl}\tau_{kl} - \gamma_{ij}\theta . \qquad (3.12)$$

The principle may be stated in two different forms which are the perfect analogues of Green's principle and Castigliano's principle in uncoupled thermoelasticity. Define two functionals

$$\begin{aligned} I_u &= \int_V (L - F_i u_i)dV - \oint_A p_i u_i dA \\ L &= \frac{1}{2}c_{ijkl}\varepsilon_{ij}\varepsilon_{kl} - \beta_{ij}\varepsilon_{ij}\theta \end{aligned} \qquad (3.13)$$

and

$$\begin{aligned} I_\tau &= \int_V (N - F_i u_i)dV - \oint_A p_i u_i dA \\ N &= \frac{1}{2}b_{ijkl}\tau_{ij}\tau_{kl} - \gamma_{ij}\tau_{ij}\theta . \end{aligned} \qquad (3.14)$$

Then

(3.15) $$\delta I_u = 0.$$

"Among all geometrically admissible deformations $u_i^* = u_i + \delta u_i$ the one u_i actually occurring is characterized by a minimum value of I_u".

Similarly,

(3.16) $$\delta I_\tau = 0.$$

"Among all statically admissible states of stress $\tau_{ij}^{**} = \tau_{ij} + \delta\tau_{ij}$ the one τ_{ij} actually occurring is characterized by a minimum value of I_τ ".

The proof of the minimum properties rests on the positive definiteness of the matrices c_{ijkl} and b_{ijkl} which, in turn, is a consequence of the second law of thermodynamics ensuring positive energy dissipation. Therefore, the second variations of I_u and I_τ are positive :

$$\delta^2 I_u = \int_V \frac{1}{2} c_{ijkl} \delta\varepsilon_{ij} \delta\varepsilon_{kl} dV > 0$$
$$\delta^2 I_\tau = \int_V \frac{1}{2} h_{ijkl} \delta\tau_{ij} \delta\tau_{kl} dV > 0.$$

(3) Schapery's Principle. From the various principles derived by Schapery we select here the one for displacement u_i and entropy displacement s_i. Besides the stress-strain relation (3.11) and the geometric relation (1.7) the linearized heat equation

$$h = -T_0 s_{,i,i} = c\theta + T_0 \beta_{ij} \varepsilon_{ij} \qquad (3.17)$$

is assumed to be valid. c is the specific heat operator at constant deformation. Solving Eq. (3.17) for the temperature θ one is led to

$$\theta = -T_0 c^{-1} \gamma \qquad (3.18)$$

where γ is given by Eq. (2.22) but now represents an operator. c^{-1} is the inverse of the specific heat operator.

Schapery's principle now states : For all variations of u_i and s_i compatible with their boundary conditions the functional

$$I = \frac{1}{2} \int_V \left\{ (c_{ijkl} \varepsilon_{kl}) * \varepsilon_{ij} + T_0 [(c^{-1} \gamma) * \gamma] + T_0 \lambda_{ij} \dot{s}_i * s_j - F_i * u_i \right\} dV - \int_{A_1} \bar{p}_i * u_i dA + \int_{A_2} \bar{\theta} * s_i n_i dA \qquad (3.19)$$

takes on a stationary value,

$$\delta I = 0 . \qquad (3.20)$$

The sign $*$ denotes convolution,

$$f*g := \int_0^t f(\tau)g(t-\tau)d\tau \qquad (3.21)$$

λ_{ij} is the thermal resistivity matrix in Fourier's law,

(3.22) $$\lambda_{ij} = [a_{ij}]^{-1}$$

where

(3.23) $$q_i = -a_{ij}\theta_{,j}, \quad \theta_{,i} = -\lambda_{ij}q_j.$$

The force vector $\bar{p}_i$ is prescribed on part A_1, and temperature $\bar{\theta}$ is prescribed on part A_2 of the surface.

By carrying out the variation one finds that the Euler equations are the equilibrium equation

(3.24) $$(c_{ijkl} + T_0\beta_{ij}\beta_{kl}c^{-1})u_{i,jl} + T_0 c^{-1}\beta_{kl}s_{i,il} + F_k = 0 \quad \text{in} \quad V$$

and the heat conduction equation

(3.25) $$c\lambda_{ij}\dot{s}_i - s_{i,ij} - \beta_{kl}u_{k,lj} = 0 \quad \text{in} \quad V.$$

The natural boundary conditions are

(3.26) $$[(c_{ijkl} + T_0\beta_{ij}\beta_{kl}c^{-1})\varepsilon_{kl} + T_0 c^{-1}\beta_{ij}s_{k,k}]n_j = \bar{p}_i \quad \text{on} \quad A_1$$

(3.27) $$s_{i,i} = -\frac{c\theta}{T_0} \quad \text{on} \quad A_2.$$

In contradistinction to the principles described in (1) and (2) Scharpery's principle includes thermal coupling.

Material properties are supposed to be temperature-independent in the preceding theorems. In a more recent paper [24] <u>Schapery</u> has suggested, for the special case of a viscoelastic medium subjected to cyclic loading, two variational principles which allow for the temperature-dependence of the shear and bulk compliances.

CHAPTER IV
MAGNETOELASTICITY

Magnetization of an elastic body affects its state of stress in three ways [25]. First, due to the appearance of a couple per unit mass **M** x **H** the stress tensor is no longer symmetric. M_i is the magnetization per unit mass and H_i is the magnetic field. Second, to the mechanical body force density ϱF_i there is added a magnetic body force density. Third, to the mechanical surface force density P_i there is added a magnetic surface force density. Furthermore, the three mechanical equilibrium equations (or equations of motion) are to be supplemented by three equations of magnetic equilibrium (or equations of magnetodynamics).

A variational principle which renders these six equations together with the corresponding boundary conditions has been given by <u>Tiersten</u> [26] for the isothermal dynamic case and by <u>Brown</u> [27] for the isothermal static case. We follow Brown[*]).

---

[*]) The Giorgi system of units will be used in this and the next chapter.

Let Π be the total potential of the body,

$$\Pi = \int_m (U - F_i u_i)dm - \oint_A P_i u_i dA - \frac{\mu_0}{2}\int_m M_i H'_i dm - \mu_0 \int_m M_i H^0_i dm \qquad (4.1)$$

where the magnetic field has been split into the applied field H^0_i and the field H'_i due to the magnetization M_i : $H_i = H^0_i + H'_i$.

The first two integrals represent the mechanical energy, cf. Eq. (2.2). However, the internal energy U is now also a function of the magnetization M_i and its gradient $M_{i,\alpha}$

$$U = U(x_{i,\alpha}, M_i, M_{i,\alpha}) . \qquad (4.2)$$

$M_{i,\alpha}$ has to be included if exchange forces are to be considered. The third term is the magnetostatic self-energy, and the fourth integral is the energy in the applied field. X_i are coordinates after deformation, cf. Sec. 2.1.

To find the equilibrium conditions, we set the first variation $\delta\Pi$ of Π equal to zero for arbitrary variations $\delta x_i = \delta u_i$ and δM_i. As a side condition for a ferromagnetic material at sufficiently low temperature we may suppose that

$$M_i M_i = M_s^2(T) = \text{constant.} \qquad (4.3)$$

Combination of this saturation condition with Eq. (4.1) leads to

$$I = \Pi - \frac{1}{2}\int_V \lambda M_i M_i dV - \frac{1}{2}\oint_A \mu M_i M_i dA \qquad (4.4)$$

where the space functions λ and μ are Lagrangian multipliers.

We perform now the sequence of steps of the variation procedure.

$$\delta I = \int_m \left[\frac{\partial U}{\partial x_{i,\alpha}} \frac{\partial}{\partial X_\alpha} \delta x_i + \frac{\partial U}{\partial M_i} \delta M_i + \frac{\partial U}{\partial M_{i,\alpha}} \frac{\partial}{\partial X_\alpha} \delta M_i - F_i \delta u_i \right] dm -$$

$$- \oint_A P_i \delta u_i dA - \frac{\mu_0}{2} \delta \int_m M_i H_i' dm -$$

$$- \mu_0 \int_m \left[H_i^0 \delta M_i + M_i \delta H_i^0 \right] dm - \int_V \lambda M_i \delta M_i dV - \oint_A \mu M_i \delta M_i dA .$$

In the fourth integral, since the applied field H_i^0 does not depend on M_i, we have immediately $\delta H_i^0 = H_{i,j}^0 \delta x_j$
Now, using $dm = \varrho dV$,

$$\int_m \frac{\partial U}{\partial x_{i,\alpha}} \frac{\partial}{\partial X_\alpha} \delta x_i dm = \int_V \varrho \frac{\partial U}{\partial x_{i,\alpha}} \frac{\partial \delta x_i}{\partial x_j} x_{j,\alpha} dV = \oint_A \varrho \frac{\partial U}{\partial x_{i,\alpha}} x_{j,\alpha} n_j \delta x_i dA -$$

$$- \int_V \left(\varrho \frac{\partial U}{\partial x_{i,\alpha}} x_{j,\alpha} \right)_{,j} \delta x_i dV$$

$$\int_m \frac{\partial U}{\partial M_{i,\alpha}} \frac{\partial}{\partial X_\alpha} \delta M_i dm = \int_V \varrho \frac{\partial U}{\partial M_{i,\alpha}} x_{j,\alpha} \frac{\partial \delta M_i}{\partial x_j} dV =$$

$$= \oint_A \varrho \frac{\partial U}{\partial M_{i,\alpha}} x_{j,\alpha} n_j \delta M_i \, dA - \int_V \left(\varrho \frac{\partial U}{\partial M_{i,\alpha}} x_{j,\alpha}\right)_{,j} \delta M_i \, dV .$$

The variation of the magnetic self-energy is difficult. We follow <u>Brown</u> and derive first the expression for the time rate of the energy. Employing the equations, for an arbitrary scalar function $F(x,t)$,

$$\frac{d}{dt}\int_m F \, dm = \int_m \frac{dF}{dt} \, dm = \int_m \left(\frac{\partial F}{\partial t} + v_i F_{,i}\right) dm , \qquad v_i = \dot{x}_i := \left(\frac{\partial x_i}{\partial t}\right)_x$$

and

$$\frac{d}{dt}\int_m F \, dm = \int_V \left[\frac{\partial}{\partial t}(\varrho F) dV + (\varrho F v_i)_{,i}\right] dV$$

and making use of the reciprocity relation, cf. [25]

$$\int_V \varrho M_i \frac{\partial H'_i}{\partial t} \, dV = \int_V H'_i \frac{\partial(\varrho M_i)}{\partial t} \, dV$$

we have :

$$\frac{d}{dt}\int_m M_i H'_i \, dm = \int_m \left[H'_i \frac{dM_i}{dt} + M_i \left(\frac{\partial H'_i}{\partial t} + v_j H'_{i,j}\right)\right] dm \qquad (a)$$

and

(b)
$$\frac{d}{dt}\int_m M_i H'_i \, dm = \int_V \left[\frac{\partial(\varrho M_i)}{\partial t} H'_i + \varrho M_i \frac{\partial H'_i}{\partial t}\right] dV + \int_V (\varrho M_i H'_i v_j)_{,j} \, dV =$$
$$= 2\int_m M_i \frac{\partial H'_i}{\partial t} \, dm + \int_V (\varrho M_i H'_i v_j)_{,j} \, dV.$$

Subtracting Eq. (b) from twice Eq. (a), we get

(c)
$$\frac{d}{dt}\int_m M_i H'_i \, dm = 2\int_m \left[H'_i \frac{dM_i}{dt} + M_i v_j H'_{i,j}\right] dm - \int_V (\varrho M_i H'_i v_j)_{,j} \, dV.$$

Now, H'_i and M_i are discontinuous on the surface A of the body and the product $\varrho M_i H'_i$ suffers a jump, [25] p.67, equal to

$$[\varrho M_i H'_i] = m_n^2$$

where $m_i = \varrho M_i$ and m_n is the component in the direction of the surface normal as the surface is approached from the inside. In transforming the last integral on the rhs of Eq. (c) into a surface integral, Gauss'formula, therefore, has to be replaced by *)

(d)
$$\int_V (\varrho M_i H'_i v_j)_{,j} \, dV = -\oint_A [\varrho M_i H'_i] v_j n_j \, dA = -\oint m_n^2 v_i n_i \, dA.$$

---

*) cf. [28], p. 427.

Eq. (c) then goes over into

$$\frac{d}{dt}\int_m M_i H'_i \, dm = 2\int_m \left[H'_i \frac{dM_i}{dt} + M_i v_j H'_{i,j}\right] dm + \oint_A m_n^2 v_i n_i \, dA . \quad (4.5)$$

Since $\nabla \times H' = 0$ or $H'_{i,j} = H'_{j,i}$ from Maxwell's equations, $v_j H'_{i,j} = v_j H'_{j,i}$ on the rhs.

The fundamental equation (4.5) can now be used to obtain the variation of the magnetic self-energy. One simply has to replace dM_i/dt and v_i by δM_i and δx_i, respectively. This leads to

$$-\frac{\mu_0}{2}\delta\int_m M_i H'_i \, dm = -\mu_0\int_m \left[H'_i \delta M_i + M_i H'_{j,i} \delta x_j\right] dm - \frac{\mu_0}{2}\oint_A m_n^2 n_i \delta x_i \, dA .$$

If we now combine the preceding results we get

$$\delta I = \int_V \left\{\left[-\left(\varrho \frac{\partial U}{\partial x_{i,\alpha}} x_{j,\alpha}\right)_{,i} - \mu_0 \varrho M_j H_{i,j} - \varrho F_i\right]\delta x_i + \right.$$

$$+ \left[\varrho\frac{\partial U}{\partial M_i} - \left(\varrho\frac{\partial U}{\partial M_{i,\alpha}} x_{j,\alpha}\right)_{,i} - \mu_0\varrho H_i - \lambda M_i\right]\delta M_i \bigg\} dV +$$

$$+ \oint_A \left\{\left[\varrho\frac{\partial U}{\partial x_{i,\alpha}} x_{j,\alpha} n_j - \frac{\mu_0}{2} m_n^2 n_i - P_i\right]\delta x_i + \right.$$

$$+ \left[\varrho\frac{\partial U}{\partial M_{i,\alpha}} x_{j,\alpha} n_j - \mu M_i\right]\delta M_i \bigg\} dA = 0 .$$

Therefore, the following equations and boundary conditions are obtained.

Mechanical equilibrium :

(4.6)
$$\left(\varrho \frac{\partial U}{\partial x_{i,\alpha}} x_{j,\alpha}\right)_{,j} + \mu_0 \varrho M_j H_{i,j} + \varrho F_i = 0 \quad \text{in} \quad V$$

$$\varrho \frac{\partial}{\partial x_{i,\alpha}} x_{j,\alpha} n_j - \frac{\mu_0}{2} m_n^2 n_i - P_i = 0 \quad \text{on} \quad A.$$

Magnetic equilibrium :

(4.7)
$$\varrho \frac{\partial U}{\partial M_i} - \left(\varrho \frac{\partial U}{\partial M_{i,\alpha}} x_{j,\alpha}\right)_{,j} - \mu_0 \varrho H_i - \lambda M_i = 0 \quad \text{in} \quad V$$

$$\varrho \frac{\partial U}{\partial M_{i,\alpha}} x_{j,\alpha} n_j - \mu M_i = 0 \quad \text{on} \quad A.$$

The tensor

(4.8)
$$\tau_{ij} = \varrho \frac{\partial U}{\partial x_{i,\alpha}} x_{j,\alpha}$$

is the well-known Cauchy stress tensor. The equations of mechanical equilibrium may therefore be written as

(4.9)
$$\tau_{ij,j} + \mu_0 \varrho M_j H_{i,j} + \varrho F_i = 0 \quad \text{in} \quad V$$

$$\left(\tau_{ij} - \frac{\mu_0}{2} m_n^2 \delta_{ij}\right) n_j = P_i \quad \text{on} \quad A.$$

The vector

$$\tilde{H}_i = H_i - \frac{1}{\mu_0}\left[\frac{\partial U}{\partial M_i} - \frac{1}{\varrho}\left(\varrho\frac{\partial U}{\partial M_{i,\alpha}}x_{j,\alpha}\right)_{,j}\right] \qquad (4.10)$$

appearing in Eq.(4.7)$_1$ is the "effective field" which for equilibrium must be in (or opposite to) the direction of M_i. Indeed, upon eliminating λ and μ from Eqs. (4.7) by multiplying by M_k and taking the antisymmetric part

$$-\mu_0\varrho(\tilde{H}_iM_k - \tilde{H}_kM_i) - \lambda(M_iM_k - M_kM_i) = 0$$

we get

$$\begin{aligned}\tilde{H}_iM_k - \tilde{H}_kM_i &= 0 \quad \text{or} \quad \tilde{\mathbf{H}}\times\mathbf{M} = 0 \quad \text{in } V \\ \left(\frac{\partial U}{\partial M_{i,\alpha}}M_k - \frac{\partial U}{\partial M_{k,\alpha}}M_i\right)x_{j,\alpha}n_j &= 0 \quad \text{on } A.\end{aligned} \qquad (4.11)$$

$\tilde{\mathbf{H}}\times\mathbf{M}$ is the total torque per unit volume which must vanish.

It should be pointed out that U as a function of $x_{i,}$, M_i and $M_{i,\alpha}$ does not satisfy the principle of frame indifference (principle of objectivity). In transforming U to make it objective one may, for instance, use the variables $e_{\alpha\beta}$ and <u>Toupin's</u> vector $\Pi_\alpha = M_i x_{i,\alpha}$. The resulting equations are complicated and will not be given here.

If deformations are small and if the applied field H_i^0 is large compared to H_i' the preceding equations may be linearized. This leads to drastic simplifications, particularly so if, in addition, exchange forces are neglected. It is then not difficult to extend the principle to the nonisothermal case since equation (3.17) of linearized heat conduction remains unaffected by the presence of the magnetic field. A corresponding variational principle has been given by Nowacki [29]. It includes also the effect of electric current conduction. The quantities to be varied are the mechanical displacement u_i, the entropy displacement s_i and a certain nonsymmetrical tensor Φ_{ij} modelled after s_i and defined by $\dot{\Phi}_{ij} = -H_{j,i}$. The results of the principle are, however, at variance with those obtained by linearizing the preceding equations. This is due to the different magnetic model used in the derivation of the principle.

CHAPTER V
PIEZOELECTRICITY

The constitutive equations of a polarizable (but not magnetizable) dielectric may be written as

$$\tau_{ij} = \frac{\partial U}{\partial \varepsilon_{ij}}, \quad E_i = \frac{\partial U}{\partial D_i}. \qquad (5.1)$$

Here, $U = U(\varepsilon_{ij}, D_i)$ is the internal energy, depending on infinitesimal strain ε_{ij} and electric displacement D_i. The vector E_i is the electric field intensity. As is well-known the latter may be expressed in terms of a vector and scalar potential $\mathbf{A}$ and φ, respectively, as

$$E_i = -\varphi_{,i} - \dot{A}_i. \qquad (5.2)$$

The basic simplifying assumption of piezoelectricity, [30], now is that

$$|\dot{A}_i| \ll |\varphi_{,i}|. \qquad (5.3)$$

Then,

$$E_i = -\varphi_{,i}. \qquad (5.4)$$

Under the additional assumptions of linear, isothermal constitutive equations, absence of volume charges and linearized equation of motion (i.e., neglect of electric volume forces), <u>Tiersten</u> [30] has formulated a simple extension of Hamilton's principle as follows.

With the aid of a Legendre transform the electric enthalpy, $H(\varepsilon_{ij}, E_i)$ is introduced as

(5.5) $$H(\varepsilon_{ij}, E_i) = U(\varepsilon_{ij}, D_i) - E_i D_i .$$

Then, using (5.1),

(5.6) $$\tau_{ij} = \frac{\partial H}{\partial \varepsilon_{ij}}, \quad D_i = -\frac{\partial H}{\partial E_i} .$$

Also, from Maxwell's equations,

(5.7) $$D_{i,i} = 0 .$$

Since a linear material is assumed H will have the form

(5.8) $$H = \frac{1}{2} c_{ijkl} \varepsilon_{ij} \varepsilon_{kl} - e_{ijk} E_i \varepsilon_{jk} - \frac{1}{2} b_{ij} E_i E_j .$$

The 3 equations (5.4), (5.6) and (5.7) together with the 3 equations of motion

(5.9) $$\tau_{ij,i} + \varrho F_j = \varrho \ddot{u}_j$$

and the 6 kinematic relations (1.7) constitute a set of altogether 22 equations in 22 variables. They can readily be reduced to 4 equations in the 4 variables u_i and φ :

(5.10a) $$c_{ijkl} u_{k,li} + e_{kij} \varphi_{,ki} = \varrho \ddot{u}_j$$

Tiersten's Principle

$$e_{ikl} u_{k,li} - b_{ik} \varphi_{,ki} = 0 . \tag{5.10b}$$

The corresponding boundary conditions specify any combination of 4 quantities out of $p_i = \tau_{ij} n_j, u_i, D_i n_i, \varphi_{,i}$.

Tiersten's variational principle now states that

$$\delta \int_{t_1}^{t_2} (K - \Pi) dt = 0 \tag{5.11}$$

where

$$K = \frac{1}{2} \int_V \varrho \dot{u}_i \dot{u}_i dV$$

$$\Pi = \int_V \varrho (H - F_i u_i) dV - \oint_A (p_i u_i - \sigma \varphi) dA \tag{5.12}$$

σ is the electric charge on the surface A.

The quantities to be varied in Eq. (5.11) are u_i and E_i, with Eqs. (1.7), (5.4) and (5.6) to be satisfied. The variation procedure follows the usual pattern and leads to

$$\delta \int_{t_1}^{t_2} dt \left\{ \int_V [(\tau_{ij,i} + \varrho F_j - \varrho \ddot{u}_j) \delta u_j + D_{i,i} \delta \varphi] dV + \right.$$

$$\left. + \oint_A [(p_i - \tau_{ij} n_j) \delta u_i - (\sigma + n_i D_i) \delta \varphi] dA \right\} = 0 . \tag{5.13}$$

Hence, Eqs. (5.7) and (5.9) are recovered together with the boundary conditions

$$\tau_{ij} n_j = p_i \qquad \text{on that part of A where } u_i \text{ is not prescribed}$$

and

$$D_i n_i = -\sigma \qquad \text{on that part of A where } \varphi \text{ is not prescribed.}$$

The surface charge may be taken as zero on all surfaces on which it much be prescribed.

The principle in the form (5.13) has been used by <u>Tiersten</u> [30] to obtain approximate solutions to the problem of vibrations of piezoelectric plates.

REFERENCES

[1] D. Bernoulli : Correspondance math. et phys. Letters to Euler dated May 24 and November 8, 1738.

[2] G. Green : On the laws of reflexion and refraction of light at the common surface of two noncrystallized media. Cambr. Phil. Soc. Trans.7 (1838),1.

[3] A. Castigliano : Théorie de l'équilibre des systèmes élastiques. Turin 1879.

[4] G. Kirchhoff : Mechanik. 11. und 28. Vorlesung. B.G. Teubner, Leipzig 1876.

[5] E. Reissner : On a variational theorem in elasticity. J. Math. Phys. 29 (1950), 90.

[6] E. Reissner : On a variational theorem for finite elastic deformations. J. Math. Phys. 32 (1953), 129.

[7] H. Parkus : Thermoelasticity. Blaisdell Publishing Comp., Waltham-Toronto-London 1968.

[8] R. Trostel : Genaherte Berechnung von Warmespannungen mit Hilfe der Variationsprinzipien der Elastostatik. Ing. Arch. 29 (1960), 388.

[9] H. Parkus : Ueber eine Erweiterung des Hamilton'schen Prinzipes auf thermoelastische Vorgange. Federhofer-Girkmann Festschrift Wien 1950. Verlag F. Deuticke, p. 295.

[10] M.A. Biot : Thermoelasticity and irreversible thermodynamics. J.Appl.Physics 27 (1956), 240.

[11] G. Herrmann : On variational principles in thermoelasticity and heat conduction. Quart.Appl.Math. 21 (1963), 151.

[12] M. Ben-Amoz : On a variational theorem in coupled thermoelasticity. J. Appl. Mech. 32 (1965), 943.

[13] R.E. Nickell and J.J. Sackman : Variational principles for linear coupled thermoelasticity. Quart.Appl.Math. 26 (1968), 11.

[14] P. Rafalski : A variational principle for the coupled thermoelastic problem. Int.J.Engng Sci. 6 (1968), 465.

[15] G. Herrmann : On a complementary energy principle in linear thermoelasticity. J.Aero/Space Sc.25 (1958), 660.

[16] J.L. Zeman : A method for the solution of stochastic problems in linear thermoelasticity and heat conduction. Int.J. Solids Structures 2 (1966), 581.

[17] A. Wang and W. Prager : Thermal and creep effects in work-hardening elastic-plastic solids. J. Aer.Sci. 21 (1954), 581.

[18] J.L. Sanders, Jr.,H.G. McComb, Jr., and F.R. Schlechte : A variational theorem for creep with applications to plates and columns. NACA-Rep. 1342, 1958.

[19] M.A. Biot : Variational principles in irreversible thermodynamics with application to viscoelasticity. Phys. Review 97 (1955), 1463.

[20] R.A. Schapery : Irreversible thermodynamics and variational principles with applications to viscoelasticity. Aeronautical Research Laboratories Report No. ARL 62-418, Wright-Patterson AFB, Ohio 1962.

[21] W. Olszak and P. Perzyna : Variational theorems in general viscoelasticity. Ing.-Arch.28 (1959), 246.

[22] P. Rafalski : Variational principle for the problem of dynamic thermal stresses in a linear viscoelastic body. Bull. Acad.Pol. Sciences. Serie sc.techn. 16 (1968), 295.

References

[23] J. Hult : Creep in Engineering Structures. Blaisdell Publishing Comp.Waltham-Toronto-London 1966. Section 5.1.

[24] R.A. Schapery : Thermomechanical behavior of viscoelastic media with variable properties subjected to cyclic loading. J.Appl. Mech. 32 (1965), 611.

[25] W.F. Brown, Jr. : Magnetoelastic Interactions. Springer-Verlag Berlin-Heidelberg-New York 1966.

[26] H.F. Tiersten : Variational principle for saturated magnetoelastic insulators. J. mathematical Physics 6 (1965), 779.

[27] W.F. Brown, Jr. : Theory of magnetoelastic effects in ferromagnetism. J.Appl.Phys. 36 (1965), 994.

[28] A.C. Eringen : Mechanics of Continua. J. Wiley and Sons, New York-London-Sidney 1967.

[29] W. Nowacki : Problem of linear coupled magneto-thermoelasticity. Bull.Acad.Polonaise Sci. 13 (1965), 331.

[30] H.F. Tiersten : Linear piezoelectric plate vibrations. Plenum Press. New York 1969.

MIX
Papier aus verantwortungsvollen Quellen
Paper from responsible sources
FSC® C105338

If you have any concerns about our products,
you can contact us on
ProductSafety@springernature.com

In case Publisher is established outside the EU,
the EU authorized representative is:
Springer Nature Customer Service Center GmbH
Europaplatz 3, 69115 Heidelberg, Germany

Printed by Libri Plureos GmbH
in Hamburg, Germany